SAY NO TO MEAT

Simple Tips and Easy
Recipe...
Out...

GW00632134

summersdale

SAY NO TO MEAT

An Hachette UK Company
www.hachette.co.uk

Summersdale Publishers Ltd
Part of Octopus Publishing Group Limited
Carmelite House
50 Victoria Embankment
LONDON
EC4Y 0DZ
UK

www.summersdale.com

Printed and bound in Malta

Printed on 100% recycled card and paper

ISBN: 978-1-78685-971-6

Substantial discounts on bulk quantities of Summersdale books are available to corporations, professional associations and other organizations. For details contact general enquiries: telephone: +44 (0) 1243 771107 or email: enquiries@summersdale.com.

Disclaimer
Neither the author nor the publisher can be held responsible for any loss or claim arising out of the use, or misuse, of the suggestions made herein.

CONTENTS

WHAT'S SO BAD ABOUT MEAT?

It's never been easier to reduce the amount of meat and fish you consume. It's also never been more important that you do so. Gone are the days of nut roasts and limited menu options. We are enjoying a meat-free heyday, blessed by a cornucopia of supermarket meat alternatives and vegetarian restaurants.

The world has changed and the way we produce meat has changed too. We are no longer hunting our food or even farming on a local level. The animal agriculture business is global and its consumption is massive. Animal agriculture gobbles up land and water and produces enormous quantities of greenhouse gases, causing serious strain on our environment. The United Nations has projected that our global population will have swollen to 9 billion by 2050 and that we could need three planets' worth of resources to maintain that population with our current lifestyles.

With the massive technological progress made in food production, including the ability to create alternative meat products that look and taste like animal meat, many people believe that it isn't ethically acceptable to kill animals for food. If you have the option to eat healthily and sustainably and *not* kill an animal, doesn't it make sense to take it?

Whatever your reason, this book contains masses of information to help you take the first steps towards cutting meat from your diet. Look for the symbols below to see how each tip can help you.

KEY

	This is an inexpensive way to cut out animal products
	This option involves a new way to do things
	This tip includes a meat replacement idea
	This option reduces the amount of meat you consume
	This option is for those who have mastered the basics

Some types of meat carry health risks when consumed in large quantities. Studies have linked eating a lot of processed and red meat with an increased chance of developing bowel cancer. Around 34,000 cancer deaths a year can be linked to diets rich in processed meat.

Studies show that people who follow a plant-based diet can experience health benefits including a lower risk of heart disease, lower cholesterol levels and a reduced risk of some types of prostate and breast cancer.

Some types of meat are more environmentally friendly than others. Gram for gram, chicken farming produces three to ten times less greenhouse gas emissions than beef, in part due to the large rate of methane emissions from cows. If you're looking to reduce your impact on the environment, but are scared of going totally meat free all in one go, you can start making a difference by swapping your beef burger for chicken.

MEAT CONSUMPTION HAS ROCKETED OVER RECENT DECADES. THE WORLDWIDE CONSUMPTION OF MEAT IN 1960 WAS AROUND 7 MILLION METRIC TONS. THE GLOBAL CONSUMPTION OF MEAT TODAY IS CLOSER TO 30 MILLION METRIC TONS.

Animal agriculture consumes a vast amount of water. It can take up to 683 gallons of water to produce 1 gallon of milk, and 2,400 gallons of water to produce 1 pound of beef. By comparison it takes only 244 gallons of water to produce 1 pound of tofu, a traditional Asian food made from soy beans that has become a popular vegan meat alternative.

THE GLOBAL ANIMAL AGRICULTURAL BUSINESS PRODUCES MORE GREENHOUSE GASES THAN THE WORLD'S CARS, BUSES, PLANES, TRAINS AND OTHER TRANSPORTATION SYSTEMS COMBINED.

According to the United Nations nearly 90 per cent of the world's fishing spots are overfished. Even farmed fish feed on wild fish caught from the sea – it takes over a kilo of wild fish to produce only 400 g of salmon.

THE NUMBER OF FISH IN THE OCEAN HAS REDUCED BY HALF SINCE 1970 AND AROUND 80 PER CENT OF COMMERCIAL FISH SPECIES HAVE DISAPPEARED AS A RESULT OF OVERFISHING.

COMMERCIAL FISHING IS ALSO RESPONSIBLE FOR THE DEATH OF THE MANY ANIMALS IT CATCHES ACCIDENTALLY, SUCH AS DOLPHINS, TURTLES AND SHARKS.

Bottom trawling – the commercial fishing technique of dragging heavily weighted nets across the ocean floor – is responsible for much of fishing's destruction of marine environments. Modern heavyweight trawler nets destroy thriving coral beds, depriving ocean creatures of their habitat. In some areas, such as the coral seamounts off Southern Australia, nearly 90 per cent of the coral beds are reduced to rock.

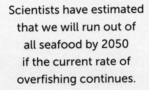

Scientists have estimated
that we will run out of
all seafood by 2050
if the current rate of
overfishing continues.

WHERE TO START

Well done! You've taken the first step by deciding to make a positive choice in your life and reduce your meat consumption. So what's next? It's OK to start small and make just one change – in fact that can be the best way to form lasting habits. Here are a few methods and techniques to help you reduce your meat consumption in an easy and sustainable way.

1

MEAT FREE MONDAYS

Meat Free Monday is a global campaign to encourage people to cut meat from their diets for just one day of the week. Launched by former Beatle Paul McCartney in 2009, the campaign aims to get consumers thinking about how cutting meat can improve their health, help them save money and be more environmentally friendly – and to show them how easy it is to do! Its informative website www.meatfreemondays.com offers great information and recipe ideas. Participating in Meat Free Mondays is a great way to begin cutting meat from your diet, and at only one day a week it's a nice, achievable start. Search the hashtag #meatfreemondays on social media and you'll find a thriving community chatting, sharing tips and recipes and pictures of their food.

2

EAT MEAT ONCE A DAY

Real meat-a-holics might want to start by slowing their meat intake before they start making serious cuts. The World Health Organization recommends cutting your processed meat intake to reduce the risk of some cancers and has recognized links between red meat consumption and high cholesterol and heart disease. Start by reducing the number of meals containing meat to one a day. Opt for a bacon sandwich for breakfast OR a chicken fajita wrap for lunch OR lamb curry for dinner rather than all of these meat-feasts in one day. The gradual transition will mean that you're not changing all your eating habits at once and you'll be able to acquire healthy habits and delicious recipes over a time that's manageable for you.

3

DON'T EAT MEAT JUST FOR THE SAKE OF IT

Are you truly enjoying all the meaty meals that you're eating, or are you eating some because that's just what you know? It's hard to keep coming up with fresh ideas and so it's easy to find yourself eating meat-based meals simply out of habit. Think about the less inspiring meals that you eat – the ones you don't *love* – and try replacing them with meat-free recipes and products. Perhaps you always buy the same ham sandwich at lunch – spend your next week experimenting with new lunchtime meals. Perhaps there's a meaty pasta dish that's your go-to when you need to make a quick dinner – try replacing the meat with some roasted red peppers or artichoke hearts instead. You'll barely notice that you're reducing your meat intake!

4

TRY 3-2-1 (MEATY MEALS A WEEK)

Reduce the number of meat-based meals that you eat in a week in stages. Start by cutting the number of meaty meals down to three a week. Try that for a few weeks – or months – and if it is working well for you, and you are feeling energetic and motivated, cut down again to two a week. You can repeat the process to get to one meaty meal a week and then, if you like, none at all. Remember that this is a technique to help you, not a binding contract, so you can increase the number of meaty meals you eat at any time to suit you, and you're allowed to make exceptions for special occasions or just because you want to!

5

KEEP UP YOUR NUTRITIONAL INTAKE

A meat-based diet will provide you with protein, iron and B12, so when you start to move away from meat remember that your body still needs these nutrients. Eating less meat means there's more space for delicious and healthy alternatives and lots more nutrient-rich foods that should help you avoid deficiencies or a drop in energy. Dark leafy greens such as chard, spinach or kale are rich in protein, iron and B12; try bulking out your meals with these veggies if you notice your energy levels flagging. See "A Balanced Diet" on p.35 for more information.

6

ENERGY-MAINTAINING REPLACEMENTS: CEREALS AND GRAINS

Cereals and grains are a family of foods that include rice, wheat, barley, oats and maize and their derivatives, such as flour. They are a good source of protein and energy and are widely available. As such, a good way of supplementing your diet when decreasing your meat portion is to start by increasing your cereal or grain portion. Help yourself to an extra spoonful of rice with your vegetable curry and bulk out your lunchtime salads with some quinoa or couscous.

ENERGY-MAINTAINING REPLACEMENTS: LEGUMES

It's a common joke that a vegetarian diet is full of beans and, frankly, that's a fair point. But that doesn't mean you'll be forever eating nothing but baked beans. Increasing your intake of legumes – including lentils, chickpeas, black beans, kidney beans, peas, tofu products (a soy bean derivative) and alternative meat products made from legumes – is an excellent and varied way to consume the nutrients that you would usually get from the meat in your diet. Legumes are rich in plant protein, iron, calcium and zinc – all usually found in meat – as well as B vitamins, potassium and folate. And they're low in fat, too! Fantastic!

8

ENERGY-MAINTAINING REPLACEMENTS: NUTS

You may feel a little hungrier between meals once you've started to cut meat from your diet. Nuts are rich in protein and fats so it only takes a small handful of them to make you feel full and energized. Almonds, pistachios and hazelnuts all contain unsaturated fats, which release energy slowly and keep your energy levels stable, unlike high-sugar snacks such as chocolate or cake. You can also bulk out your recipes with nuts to make your meals more satisfying – try sprinkling pine nuts on top of a risotto, adding a handful of walnuts to a salad, or adding peanuts to a curry for extra flavour and nutrients.

9

ENERGY-MAINTAINING REPLACEMENTS: VEGGIES

There is no better time to try out new and exciting vegetables than when you are cutting meat from your diet. Vegetables can fall by the wayside when meat is on the scene, often served with little preparation and less seasoning. Now is the time to branch out and try new things – there's a whole world of delicious vegetables out there, they just need to be prepared and seasoned well. Try roasting, honey-glazing, sautéing or grilling veggies to bring out their natural flavours and sweetness, and aim to eat a variety of colours of vegetables to get the full rainbow of nutrients – you won't believe what you were missing out on!

10

ENERGY-MAINTAINING REPLACEMENTS: FRUIT

One of the amazing benefits of cutting out meat is that it's a brilliant opportunity to eat more healthily – your body will thank you for adding lots more nutritious foods and for eating a bigger variety of natural foodstuffs. Fruit makes another great snack between meals or a delicious dessert, and increasing your intake will also help enormously in your efforts to reach your eight a day quota. Bear in mind that the high sugar levels in some fruits can cause energy levels to peak and then dip, but it's still a much healthier alternative than reaching for a chocolate bar or cake.

11

NEW RECIPE, NEW YOU

Now is the perfect time to experiment with new cuisines and new recipes. While you can simply swap meat in your favourite recipes with meat replacements when you fancy some home comforts, you could also try your hand at a few new recipes. Be inspired by cuisines from around the world, especially ones where meat is not the mainstay of a dish, such as in South Asian or East African food – why not try adding a Japanese, Ethiopian or Bengali dish to your repertoire?

12

MEAL PLANNING

It will be a lot easier to avoid cooking fatigue, which can lead to you falling back into old habits and opting for a familiar meaty meal, if you have planned and shopped for all your mid-week meals at the weekend. Sticking to a menu will help avoid fresh ingredients going off, and will save you time and money – especially if you bulk-buy the staple ingredients and then use a few of the same fresh ingredients in different ways and different meals throughout the week for variety. Thinking about what you're going to cook for the week ahead will also help you find plenty of delicious new recipes to try, and keep you motivated to continue.

13

MEAL PREPPING

Go one step further than planning your meals and spend your Sunday afternoon batch-cooking a couple of different, exciting and delicious lunches and dinners for the week ahead. You could even batch-roast root or Mediterranean veg for a versatile base for a variety of appealing meals. Most meals will last 3–5 days when stored in an airtight container in the fridge and for up to three months in the freezer. This will vary depending on the ingredients you use – always check guidelines online before storing your food. Let your meat cravings do battle against several days' worth of pre-prepared, mouth-watering food – they'll soon be vanquished.

14

SAVE YOUR MEAT MONEY

You could save some serious cash by cutting or reducing your meat intake. A study published in the *Journal of Hunger and Environmental Nutrition* found that, on average, non-meat eaters spent $750/£565 less a year on food than their meat-eating counterparts. Every time you go shopping, tot up how much money you'd usually spend on meat products weekly and pop that in a jar when you get home. You could keep it for a rainy day or treat yourself to something wonderful.

15

PERCEPTION OF MEAT

Cutting meat from your diet is a lot easier once you've changed your mentality about it. Consider the following facts: you can get all the nutrients that you find in meat from plant and dairy sources, so meat is not essential to your well-being. Eating too much of certain types of meat can have negative effects on health (see pp.6–7) and carries a higher risk of food poisoning than most other food. Intensive animal agriculture is unquestionably cruel to animals and to the environment, wasting precious water and land resources and pumping out damaging greenhouse gases. Meat is also more expensive than plant-based products. That's a long list of negatives – is it really worth it?

16

GET LOCAL

Pique your interest in ingredients and support local businesses by shopping at farmers' markets and pick-your-own fruit and veg farms. There will be more variety than in the grocery section of your local supermarket and you're likely to walk away with a few bargains too. Plus, if you're cutting meat for environmental reasons, buying locally grown produce is another way to reduce the carbon footprint of your food. You can even transform your big weekend shop into an afternoon out by visiting a pick-your-own farm. Just remember to clear your freezer before setting off – you're guaranteed to come home with more exciting produce than you can get through!

17

SMALL CUTS

You can start very slowly by simply reducing the amount of meat you consume per meal, before you start cutting it out entirely. This is a useful technique for the real meat feasters who feel that cutting out meat entirely might be a bit of a shock to the system! Opt for a fry-up with sausages *or* bacon and sub in some extra mushrooms or tomatoes, or enjoy the meat of your choice for a roast dinner but add vegetarian stuffing to the plate. Every change is a valuable change and remember – slow and steady wins the race.

18

YOU'VE ALREADY STARTED!

You're likely to have a lot of the tools for giving up meat already. You don't need to change everything all at once. Just think – what meals do you already know how to cook without meat? Use these as the foundation of your new cooking regime. The next question is which of your recipes have meat as an afterthought or an "also ran"? These could be big-pot dishes with lots of delicious ingredients that equally contribute to the taste and experience of a meal, such as stir-fries, tomato-based pastas or some curries. These recipes will be easy to adapt – simply drop the meat and increase the other ingredients.

A BALANCED DIET

You may be cutting meat for many ethical reasons but your first priority should always be your personal health. Remember that meat is a good source of nutrients such as protein and iron and that you can't simply go without them – but they don't have to come from meat. This chapter will guide you through the nutrients that you will need to replace and the best foods to help you replace them.

19

PROTEIN

"How will I get my protein?" is one of the main questions on everybody's lips when they cut meat from their diet. Actually, there are a lot of foods available to vegetarians that are rich in protein. Adults are recommended to eat 0.75 g protein a day per kilogram they weigh; for a quick guide to your intake, it's estimated at around 55 g a day for an average man and 45 g for an average woman. Great vegetarian sources of protein include peanuts, sunflower seeds, almonds, black beans, chickpeas, soy protein, wild rice, quinoa, eggs, tofu, kidney beans, plain yoghurt, peas and kale.

20

IRON

The recommended daily amount of iron in your diet is around 0.08 g a day for men and 0.15 g for women between the ages of 19–50 (post-menopausal women require less iron after their periods stop). Surprisingly, cooking in cast-iron pots, especially acidic foods such as tomato sauce, can increase your intake of iron. Otherwise you can get your iron from a variety of foods such as kidney beans, lentils, almonds, broccoli and dried apricots as well as many other veggies, nuts and seeds. You can also get some from fortified cereals such as Weetabix.

21

VITAMIN B12

B12 is essential to keep our metabolism strong and our brains working well. Tablets are available but should supplement a B12-rich diet – only 10 mg of every 500 mg of B12 is absorbed when taken in tablet form. B12 can be found in dark leafy greens such as spinach and chard or fortified non-dairy milks, cereals and some meat replacements. Marmite-loving vegetarians, this is your moment – fortified yeast extracts such as that delicious dark spread are rich in B12. You can also include nutritional yeast in your cooking for when you want a cheesy flavour without all the cheese.

22

ZINC

It's easy to overlook zinc intake when you're trying to balance your meals but it shouldn't be discounted. Although you are unlikely to think "I'm feeling a little unwell, must be my zinc deficiency", that may be the reason; if you are deficient in zinc you are more likely to be susceptible to common illnesses such as colds and the flu. You may also suffer some hair loss and increased acne. Build your zinc levels back up with tofu, black beans, kidney beans, lentils, pumpkin seeds, cashews and peanuts.

23

CALCIUM

Calcium is good for keeping your bones and teeth strong but adults don't need a great deal to be healthy. You only need to consume 0.7 g calcium a day to fulfil your nutritional requirements. It's common knowledge that milk and cheese are good sources of calcium but if you're flirting with the idea of omitting dairy products from your diet, you'll be glad to know that there are plenty of other great sources out there. These include dried figs, almonds, tofu, broccoli, chickpeas and French beans.

24

VITAMIN D

While it's true you get vitamin D from being exposed to sunlight, if you live in a cooler country it is likely that you're suffering from a vitamin D deficiency in the winter months. Taking vitamin D tablets does help replenish your stores but you may want to ensure that you are also doing the best you can with your diet. Breakfast foods are great sources of vitamin D – egg yolks, fortified cereals and plant-based milk all contain this important nutrient.

25

OMEGA-3 FATTY ACIDS

Omnivores usually turn to oily fish to get their omega-3 fatty acids. These acids help regulate cholesterol and aid the healthy development of cell membranes. Although you may have chosen to avoid fish, you can still obtain your omega-3 fatty acids from oils and some nuts, beans and fortified dairy and soy products. There is no recommended daily intake for omega-3, although you can refer to the recommendations for your daily saturated fat intake and ensure omega-3 fatty acids constitute a portion of that. Look to rapeseed and vegetable oil, flaxseed oil, walnuts and Brussels sprouts as good sources of omega-3 fatty acids.

26

EGGS

Eggs are a great source of protein as well as vitamins D and B12 and minerals such as zinc and iron. They are a good source of energy – one medium egg contains around 80 calories – and they go well with any meal. If your main motivator in cutting down on meat is your health, then eggs are the perfect substitute ingredient to maintain a healthy balanced diet. Their richness and versatility means that they fill you up *and* replace the essential nutrients you may have been missing. They can be easily added to a meal – ideas include poaching them in your curries, stirring them into fried rice and soft boiling them and adding to a salad.

27

KALE

People reducing their meat intake have to be careful about consuming enough iron. Iron deficiency, or anaemia, can lead to tiredness and fatigue, shortness of breath, cold hands and feet and brittle nails. If you've cut meat from your diet and you're noticing an increase of these symptoms then you may be suffering from an iron deficiency. Spinach is the best-known plant-based source of iron but actually, gram for gram, kale is richest in nutrients. It is high in fibre and contains good quantities of iron and vitamin C. It's very flexible and can be used in most recipes that call for spinach or simply added to most meals for the last 2–3 minutes of cooking, either in a pan or in the oven.

28

SMOOTHIES

Making dietary changes can be overwhelming, especially if you're not someone who loves cooking. If you're struggling to keep up with all your essential nutrients and replacements then consider adding a smoothie to your daily diet. Smoothies are a great way to get all your essential nutrients in one hit – try the recipe on p.107. Berries and bananas are excellent fruits to act as the base of your smoothie and you can bulk out the nutritional value with seeds such as flaxseeds or chia and vegetables such as spinach or carrot.

MEATY ALTERNATIVES

So you can't eat meat... what can you eat? Lots! For those who want to cut meat for ethical reasons but still love the taste and texture of meat, this chapter explores all the top veggie meat options; what's available, what they're good for, and what you can do with them. For those who have never been that keen on meat anyway, there are plenty of options for you too, such as tofu, aubergine and mushrooms – all excellent ways to bulk up a meat-free meal.

29

TRY MINCE!

Vegetarian mince is a great meat substitute to try when you're just starting out; you don't notice the difference when it's paired with sauces such as in a bolognese or chilli, and it needs to be cooked for a similar amount of time to meat, so in most cases you can simply follow the recipe instructions as if nothing has changed. There are plenty of brands available and the standard is pretty high across the board. The only notable difference is that you can't (easily) form vegetarian mince into burgers or meatballs. It can be a "thirsty" vegetarian meat so keep an eye on the pan; it may have absorbed more of your cooking oil or sauce than you expected.

30

TRY FAKE MEAT!

There are plenty of easy-to-cook vegetarian meat products available, with burgers, sausages and "chicken" being among the most popular and readily available. Some are made of beans and veggies encased in a crispy coating, but faux meats that taste and feel like the real thing have recently overtaken these in popularity. Vegetarian chicken nuggets are particularly close in taste and texture to their meaty counterparts, so your first steps to cutting meat can be as simple as having "chicken" nuggets and chips. There is a plethora of brands available and they can be found in both the cold aisles and the freezer aisles in stores, so you have fresh and long-life options. These products are comparable to their meaty counterparts in price, protein content and ease of cooking. Food technology has increased to the point that you can now even buy vegetarian "bleeding meat" that tastes and feels particularly authentic – although *spoiler alert* the blood is only beetroot juice. These products may not be the most nutritious food, but they do the trick if you're craving something filling and hearty. "Chicken" strips and chunks are great in curries, noodle dishes and stews – add them later than you would for real meat to avoid them absorbing too much sauce and falling apart.

31

YOU CAN EAT (FAKE) FISH TOO!

Though less commonly available than vegetarian meats, there are some substitute products for fish and other *fruits de mer* out there. The most common – and most convincing in terms of texture and flavour – are breaded or battered "white fish" fillets. These approximate the texture of flaky fish well and taste great when paired with buttery potatoes and fresh greens. Some companies offer battered "prawn" or "scampi" products; they have a good seafood flavour but their texture isn't a direct match. These work well cooked and added to a noodle dish or paella at the last minute.

32

TRY TOFU

Tofu is a good source of protein, amino acids, iron and calcium – all essential nutrients that you get from meat. Tofu's reputation as being tricky to prepare isn't exactly warranted these days, with pre-pressed or marinated tofu chunks available that need no extra preparation. If you buy standard tofu blocks packed in water, they need to be pressed before cooking but this only takes around 20 minutes and can be done while prepping your other ingredients – simply wrap in kitchen paper and place between two plates, with something heavy (such as some recipe books) stacked – carefully – on top. Tofu is perfect for throwing into a salad or rice or noodle dish for a little extra oomph, or as the base of a delicious stir-fry.

33

GET STARTED WITH SOY PRODUCTS

Most vegetarian alternatives that imitate meat's texture and flavour are made from soy "meat". This is also known as textured vegetable protein or TVP. The main ingredient of TVP is soy protein which is then shaped and textured by a machine to imitate different types of meat, and then herbs, spices and seasonings are added to create the correct flavour – this can be anything from mince to burgers to chicken to fish. As TVP's main ingredient is protein it is, unsurprisingly, high in protein and a good "like for like" substitute for meat.

Tempeh is another soy-based product, this time originating from Indonesia. Made from fermented whole soy beans pressed into a "loaf", tempeh is then sliced and used in a variety of meals including stir-fries, salads and stews. Tempeh wasn't designed specifically to replace meat but it can be used to replace the meat element in your recipe. It can even be shredded and used as an alternative to ground beef in recipes such as tacos, burritos or chillies. Like a lot of soy-based products tempeh is a good source of protein. It is also high in iron, magnesium and calcium while being low in calories and salt. Tempeh's fermentation process means it packs a lot of flavour!

34

TRY SEITAN

Seitan is a meat alternative made from wheat gluten, high in protein and low in carbs, making it a great substitute for most white or red meat. It usually has a chewy texture, similar to ribs or wings, and is very good at absorbing sauces and seasonings. The only ingredients are wheat flour and water, so it's perfect for those who are looking to cut all dairy products out of their diet, though most recipes call for extra seasonings or marinating to add flavour. It is a gluten product so sufferers of celiac disease should look elsewhere for their meat substitute (tofu is commonly gluten free).

35

GIVE MYCOPROTEIN A CHANCE

Bear with me on this one; mycoprotein is fake meat made out of fungus. Scientists in the 1960s identified a fungus species as potentially being a cheap and protein-rich foodstuff for humans. Twelve years and a lot of development later they created an effective and tasty product. Mycoprotein contains a comparable amount of protein to meat, gram for gram, but has a lower calorific value, making it a healthy alternative. Fungus-derived fake meat might not exactly sound delicious, but one of the largest meat-substitute brands on the market uses it in all its products. And they're pretty tasty! You can find mycoprotein mince, sausages, "chicken" chunks, "chicken" nuggets, "pepperoni" slices and much more!

36

FINGER ON THE PULSES

Pulses are rich in iron and protein and so are an essential ingredient for those looking to cut meat from their diets and remain healthy. They are members of the legume family and include beans, chickpeas and lentils.

Lentils soak up flavour and add bulk to recipes so they work well when used in dahls, stews or curries or as an affordable mince substitute. They are excellent additions to bolognese sauces or shepherd's pie and can even be formed into meatballs (unlike soy mince). Check out p.97 for a flavoursome lentil-based recipe. There are three main types of lentils: green, red and brown. Nutritionally they are all good for you – the real difference comes when you cook them. Red lentils have a subtler flavour and soften and disintegrate after long periods of cooking, so are perfect for thickening stews and curries. Green and brown lentils have a slightly earthy flavour, remain firm after cooking and are an ideal meat replacement.

Chickpeas are another healthy pulse that double up as an excellent, inexpensive meal bulker. Their subtle flavour works well with many cuisines and canned chickpeas are already cooked, so they can simply be drained, rinsed and popped into salads. They're a cornerstone of meat-free eating, acting as the base ingredient for vegetarian favourites hummus and falafel.

Beans such as black beans or kidney beans are low-water-consumption, low-carbon-consumption crops and are an excellent meat replacement for the environmentally friendly vegetarian. As with other pulses, including lentils and chickpeas, an 80 g serving of beans counts as one of your daily portions of fruit and veg.

37

TRY MEATY VEGGIES

Meat alternatives don't have to taste or look like meat in order to fill the space in your meal usually assigned to meat. Sometimes all you need is a large, tasty, filling item. Portobello mushrooms work really well in this capacity. Big, steak-like mushrooms that take on flavour well, portobellos can really fill you up. Pop them in a bun with a few slices of cheese, some sauce and salad and you have a perfect vegetarian burger! Oyster mushrooms' distinct meaty taste means they work well as a side dish while shiitake mushrooms' size and smoky flavour make them the perfect addition to stir-fries. Dried mushrooms have a stronger flavour than regular mushrooms and their texture when cooked is similar to that of minced meat.

Aubergines are another vegetable that can be used as a filler in meals. They are harder to cook well than mushrooms or even most alternative "meats", but are far more versatile. When sliced thickly and seared they are perfect steaks, when cooked rapidly on a high heat they are a gloriously soft and golden addition to Asian dishes and when chopped and simmered they really beef up Italian tomato-based sauces, tagines or curries. They can be breadcrumbed, battered, stuffed, caramelized, skewered... there are nearly no limits to the mighty aubergine!

38

TRY JACKFRUIT

Jackfruit is increasingly popular as a filling vegetarian option. It has long been a common ingredient in Asian cooking but Western cuisines are now picking it up as a meat substitute, and especially as an alternative to overloading on soy "meats". When young, green jackfruit is cooked it gains a chewy texture, similar to some meats. The most common product available is jackfruit "pulled pork" (check p.103 for a recipe to make your own). Jackfruit can also be added to curries and stews to imitate the texture of slow-cooked meat. Young, green jackfruit doesn't have a lot of flavour, especially when cooked, and will easily absorb the flavours of the sauce it is cooked in, which makes it great for bulking out meals healthily and adaptably.

A MEAT-FREE KITCHEN

A meat-free diet shouldn't be based on less, it should be based on *more*. More flavour, more variety, more cooking techniques, more deliciousness. Cutting meat isn't just good for your purse, for your health and for the environment, it's good for breaking out of that mealtime rut too. Stock your kitchen with a variety of herbs, spices and flavourings, and try a few of these top tips to take your meat-free cooking to the next level – you'll never miss meat again.

39

EXPERIMENT WITH CITRUS

When you think of essential ingredients for savoury cooking, oranges, lemons and limes might not be the first to leap to mind. And yet many meals can be vastly improved with a squeeze of citrus. Plenty of Asian and South American dishes, many of which are meat free or easily adapted to be meat free, use citrus as a key ingredient, and more Western dishes than you might think also use citrus fruits. Lemon pairs well with common ingredients such as spinach, soft mild cheese, olives and sage, while limes add a zing to curries, noodles, salsas and chutneys and orange is the perfect accompaniment to a sticky ginger tofu dish. Save money by using bottled lemon or lime juice for your dishes or, when using fresh lemons, keep the rinds and pickle them to use in dishes such as tagines.

40

CURRIES

Curries are ideal dishes for those taking their first steps in cutting meat from their diet. They are varied in flavour and heat, ranging from fiery tomato-based curries to smooth creamy dishes, and their punch comes from their complex mix of spices rather than the flavour of meat. Absorbent vegetables such as aubergine as well as soy meat and tofu do well to soak up the juicy curry taste. Curry recipes are incredibly diverse but you can build a good "basics" spice rack with garam masala, ground turmeric, chilli powder, medium curry powder, ground cumin and ground coriander, and keep a thumb of ginger in your fridge to add some zing to your cooking.

41

DON'T LET FRESH INGREDIENTS GO TO WASTE

Lots of Eastern cooking calls for ingredients that make a dish really stand out, such as lemongrass stalks and Thai basil, but they can be pricey to buy and often go to waste after you've made your initial dish. Prevent this by storing these ingredients in the freezer for up to three months (chopping the lemongrass first), ensuring that they won't go to waste, plus you'll have them on hand next time you want to make the dish. You can do this with more common ingredients too, such as garlic, chillies and spring onions, although they should all be chopped before freezing. You can buy lemongrass jarred and preserved in oil too, for another long-life option. Add Chinese five spice to your spice rack so you always have interesting seasoning options to hand.

42

ENHANCE DISHES WITH PASTES

A good paste can be used as a marinade, as a base for a sauce or as a sauce on its own. They are very versatile and when stored properly they have a long shelf life. Every cuisine has a good paste, from pesto to harissa to chipotle to tamarind to miso – and more! Keep a handful of pastes in the fridge for a shortcut to delicious food. For example, you could take a normal plate of sweet potato fries to the next level by coating them in chipotle paste or you could elevate a halloumi and roasted veg wrap by marinating the halloumi in harissa.

43

SEASON EVERYTHING

Seasoning your food makes a huge difference to its flavour. While it's wise to be aware of health warnings and not bury your food under a mountain of salt, a little pinch of salt here and there can really bring out the flavours in a meal. Seasoned salt is even better – some stores sell grinders packed with salt plus other tasty seasonings such as peppercorns, chilli flakes, mustard seeds, herbs, zests and spices. With these you combine all the steps this chapter recommends – adding herbs, spices, zest and other flavourings – into one easy twist of a grinder. You can buy simple combinations and flavours such as chilli salt or smoked salt and also more specific flavours such as a fajita or Szechuan mix.

Any meal can be lifted by a sprinkling of fresh herbs. Include a bunch of herbs in your weekly shop. If they are cut leafy herbs such as coriander, basil or parsley, store them in a glass filled with a little water, so they last longer. Herbs with dry stems such as rosemary and thyme can be wrapped in a damp paper towel and refrigerated. If you're not sure where to start with herbs, try adding some rosemary to a potato dish, or sprinkling some chopped basil on some pasta. It really is a very small step but it results in a huge leap in flavour.

44

OILS, VINEGARS AND SAUCES

We so often use oils as the base of our cooking that it's easy to forget that they can be a delicious garnish too. A splash of oil and balsamic vinegar really lifts a basic garden salad. A dash of chilli oil adds zing to a margarita pizza. You can elevate a speedy, basic but flavoursome noodle dish with sesame oil, soy sauce and/or rice wine vinegar. Vinegar is also an underrated ingredient and can be used for all sorts of nifty tricks. A splash of cider vinegar in the water you're boiling vegetables in can help lower your blood sugar level and even retain the vegetables' colour. Add red wine vinegar and brown sugar to onions and slow cook to caramelize them. You're just a drizzle away from a more exciting dinner!

45

REPLACE SMOKED MEATS

If you've cut cured and smoked meats such as bacon or ham you might miss that smoky flavour, but it isn't lost to you! There are all sorts of smoked cheeses available, including smoked Bavarian cheeses, smoked hard cheeses and even smoked mozzarella. You can make the base flavour of your dish smoky by cooking with smoked garlic or smoked paprika. You can also smoke your own vegetables. For example, aubergine and peppers smoke very well. Char them on the grill or a hob flame until the skin is blackened and peeling, then scrape off and enjoy the smoky insides. You can even use liquid smoke in your meal for an extra layer of flavour – add it to dried coconut to create a crispy bacon effect.

46

READY, SET, GRIDDLE!

Sometimes trying a few new cooking techniques is all it takes to help you discover the wonderful flavours and textures that veggies have to offer. Try cooking your veggies on a griddle pan for a fantastic and flavour-enhancing charring effect. Charring really brings out the sweetness in vegetables such as courgette and adds a smoky layer to tomatoes and aubergines. A cast iron pan can also add iron to your diet, as the acidity from the vegetables breaks down a little bit of the iron and it enters your food. That's a double benefit!

47

SAVE TIME SLOW COOKING

We've talked about adding more seasoning and new cooking techniques to your repertoire, but here's one way to *reduce* something: time spent cooking. Many delicious and filling meals can be created using a slow cooker, such as stews, curries, sauces and soups. Simply quickly sauté your veggies, pop the ingredients in the slow cooker at the start of the day and set it to cook. By the end of the day you'll come home and have a meal ready and waiting for you – how convenient!

48

BLENDED BASES

Another essential addition to your meat-free kitchen is a small food processor. Buy one with grating and chopping attachments and you'll wave goodbye to hours of slicing vegetables! Small food processors are also great if you find that your meat-free cooking is leading you towards Eastern cuisines. For example, Thai curries often use a paste as a base. Recipes for these can look intimidating because they contain so many ingredients but for most it's just a matter of roughly chopping the ingredients and then popping them in the blender. You can also blend roasted ingredients such as garlic, red pepper and red chillies with a little oil for an easy and delicious sauce, or whizz up some oil, lemon juice, green chilli and coriander for easy chutney.

49

BUY A SANDWICH PRESS

Upgrading your homemade sandwiches to toasties is a simple way to keep lunchtimes easy but interesting. A cheese sandwich after years of enjoying ham, chicken, bacon and other meaty fillings just isn't going to cut it. However, a melt might just do the trick! Add oregano for an Italian number, or chopped jalapeno for a spicier affair. You can branch out and toast halloumi and mint wraps for a Mediterranean feel or even make quesadillas for dinner.

50

MAKE YOUR OWN HEALTHY SNACKS

If you're feeling hungry between meals and struggling to find healthy snacks to fill the hole, a food dehydrator might be the kitchen gadget for you. You can use a food dehydrator to turn all manner of ingredients into crisps and snaps, including apples, mango, mangetout and sweet potato. Simply slice your ingredient thinly (where necessary), add to a bowl with seasoning such as salt, pepper, sugar, cinnamon or lemon, toss and lay out in the dehydrator. If you're struggling to include nutrient-rich meat replacements such as kale and chickpeas to your diet, consider dehydrating them and eating them as a tasty snack.

LIFESTYLE

Celebrations and gatherings often centre around food, and more often than not the star of the show is meat. We share meals with friends for birthdays, festivals and religious holidays. When it's hot we invite everyone over for a BBQ with all kinds of sizzling meat products and when it's cold we invite everyone over for a hearty roast dinner or a greasy fry-up. This doesn't have to change just because your diet has changed. Read on to get some top tips on navigating social occasions with a meat-free diet.

51

EATING OUT – CHOOSING YOUR CUISINE

Your first few months of cutting meat from your diet are the perfect time to explore new horizons and find new favourites. When eating out, going to the place that had that one dish that you *loved* but can't eat now will just make you sad! Try new restaurants and eateries instead and enjoy discovering dishes that you wouldn't have tried if you weren't experimenting with your diet. Eastern cuisines reliably offer a good selection of meat-free dishes whereas some gastro pubs, traditional French bistros and fine-dining establishments only offer one or two if you're lucky. Results may vary so remember to read the menu ahead of time, and check out online reviews on veggie-friendly sites.

52

EATING OUT – HOW TO READ THE MENU

Restaurants are required to label lots of different dietary requirements on their menus. Luckily for you, the most common label is "v" for vegetarian so you'll be able to select a dish easily in most eateries you attend. However, gluten free (labelled GF) is *not* the same as meat free and dishes labelled thus can and often do include meat. Keep an eye out for desserts – restaurants don't always label them, meaning that some gelatine-based desserts can slip under the radar. Panna cotta, glazed cheesecakes and jellies all run the risk of containing gelatine ("wobbly" desserts are almost guaranteed to contain it). If in doubt, don't be afraid to be "that person" who asks to check with the kitchen.

53

EATING OUT – WHEN YOU DON'T HAVE MANY OPTIONS

Try as you might, read the menu ahead as you may, suggest as many alternative restaurants as you can, sometimes you end up at a place that just doesn't have many meat-free options. Not every event – especially other people's birthdays, anniversaries etc – can be adapted to your new diet. When that happens, you may have to just go with the only meat-free main available and hope for the best. Remember to check the starters and sides sections – these often have more meat-free options than the mains, and restaurants will normally let you have a couple of these as your main meal.

54

BBQ FOOD

One of the best ways to ensure that there are veggie burgers and sausages available at the BBQ you're attending is to bring them yourself. Simple! Most veggie meats cook well on the BBQ even if they don't specify on their packaging. If you are not sure whether your alternative burger or sausage will fare well on the BBQ then wrap it in tinfoil. It will still cook well but it won't be as likely to disintegrate through the grill of the BBQ into the coals, and it protects your meat-free products from being contaminated by meaty juices on the grill.

55

FRY-UPS

Let's address those vegetarian fears head on: you don't have to give up fry-ups just because you've said goodbye to meat. If anything, they've just got better! Veggie sausages are a healthier option than – and as tasty as – meat sausages, and halloumi replaces bacon as the salty element on the plate. Besides which, all the other good stuff is happily meat free – hash browns, baked beans, fried eggs, buttered toast, fried tomatoes, fried mushrooms. Almost all cafés, from the greasiest spoon to the finest brunch spot in town, offer vegetarian breakfasts now so you're just as likely to be able to enjoy one out as you are at home.

56

WORK EVENTS

Events organized by your work can be a little trickier to navigate than those organized by your friends. If you're someone who doesn't like "making a fuss" then you may feel intimidated by your first meat-free work do. Don't be! Most workplaces, especially corporate ones, are very used to catering for people with dietary requirements, from allergies to diets that exclude certain foods for religious or ethical reasons. If you can't see any suitable food options then it's perfectly acceptable to quietly contact the organizer and ask for an option to be made available to you. Be active and involved when suggesting places to eat with colleagues so you are also providing food solutions that suit everyone. The only thing to remember is to be patient – not everyone "gets" other people's diets so be polite.

57

CHRISTMAS, EASTER AND THANKSGIVING

Sometimes an occasion calls for a really slap-up meal. For a lot of people this means meat! And not just a meat dish as a main attraction but meat as an ingredient in everything – veg cooked in lard, chopped bacon sprinkled across almost every dish and plenty of meaty side dishes for accompaniment. What's worse is that these foods are usually lovingly prepared by someone close to you, so you don't want to upset them by picking at your plate.

Just as with BBQs (p.79) the shortcut to having something you can eat at a slap-up meal is by bringing it yourself. As soon as arrangements for the big event start to be made, quietly contact the host and offer to prepare and bring your own dishes. If you feel comfortable and you know it won't add to the host's burden, you can ask them to make easy substitutions like cooking potatoes in vegetable oil instead of animal fat or lard.

The spectre of the nut roast has historically haunted many a vegetarian's festive nightmares but there are some rather tasty options available now, with delicious flavour combinations and fancy extras such as seasonal fruits like apricot or cranberry. Some brands offer festive veggie meat options, imitating the flavour of roast meat, and there are plenty of recipes for DIY veggie centrepieces. The wellington recipe on p.100 works well as a meat-free centrepiece on a special occasion.

HIDDEN ANIMAL PRODUCTS

Cutting out animal products is not as simple as merely avoiding the meat aisle at the supermarket. They have found their way into all sorts of food, so you may be consuming animal derivatives even when you least expect it! If it's important to you to avoid the meat industry altogether then you'll want to read through this chapter and learn about the foods that secretly contain meat.

58

WATCH OUT FOR GELATINE

Gelatine is a gelling agent that is made from boiled down hoof or trotter. It is a common ingredient in sweets – if a sweet is bouncy, rubbery, squishy or foamy it's very likely to contain gelatine. Set desserts such as jelly, panna cotta and some glazed, fruity cheesecakes also often contain gelatine – it even crops up as an ingredient in yoghurts more often than you'd expect. If you're worried about missing out, don't be! Veggie alternatives to gelatine (usually pectin which is derived from fruit) are available in the baking aisle of most good supermarkets and although the method of preparation is different, the results are just as good. There are lots of brands that offer gelatine-free gummy sweets and desserts, too – look in the alternative or free-from aisle in your local supermarket.

59

BE SURE ON SUET

Suet is a type of animal fat from the kidneys and loins of cows and sheep. Although it's no longer as common as it once was, you should still keep an eye out for it as it appears in some boiled, steamed or baked puddings and pastries. Stay wary at Christmas time because some traditional recipes for Christmas pudding and mincemeat include suet – check packaging for the ingredients list. Suet is also used to make tallow – another old-fashioned ingredient that still sometimes pops up in traditionally made items such as candles and soaps, and is used as a fat in some foodstuffs and in deep-frying.

60

CHECK YOUR CHEESE

Just when you thought it was safe to enter the cheese aisle! There are some cheeses that are never vegetarian because they must use rennet – an enzyme only found in calves' stomachs – in their production. These cheeses include: Parmesan, Gruyere, Pecorino, Gorgonzola, Manchego, Camembert, Grana Padano and Emmental. Some of these cheeses, such as Parmesan and Gorgonzola, must use rennet in their production to be allowed to call themselves by those names. Others, such as Camembert, are somewhat subject to this law – Camembert de Normandie always contains rennet, but vegetarian Camembert is available. Look out for "enzymes" in the ingredients list as this is often code for rennet, and only eat cheeses labelled as vegetarian.

61

LOOK OUT FOR LARD

Lard is soft pig fat usually used in cooking. Many traditional recipes for pies use lard as a binding agent in the pastry. It can also be used as a cooking agent – some restaurant chips or potatoes are cooked in lard. If a restaurant offers potatoes – particularly roast potatoes – and doesn't mark them as vegetarian on the menu then this can sometimes indicate that they're cooked in lard. If you are an avid home cook and want to use recipes that include lard, swap it out for butter or even coconut oil.

62

SHUN THE SHELLAC

You may assume that shellac, if you've heard of it at all, is a type of long-lasting nail varnish. In fact, the varnish takes its name from its key ingredient: a resin secreted from the female lac insect. The process for collecting the resin scoops up most of the bugs as well so if you're using shellac you're using the bugs' crushed bodies. Shellac is not only used in the production of nail varnish, but also in food, such as glazes, colourants and the shiny coating on most supermarket citrus fruits. If you're unsure, and the packaging or menu entry is not labelled as "suitable for vegetarians" then keep an eye out for E904 or "confectioner's glaze" in the ingredients.

63

HIDDEN FISH

Although it may be obvious, fish sauce does indeed contain fish and is not suitable for those cutting animal products from their diet. Fish sauce is a common ingredient in many Thai dishes, so always check with a restaurant when eating out – even the tofu pad thai might contain fish. Anchovies are a surprisingly common ingredient in many tasty condiments. Worcestershire sauce, olive tapenades and some salad dressings contain anchovies. Remember always to check the packaging for sneaky fish or seafood-based products: fish sauce, fish oil, cod liver oil, crustaceans, molluscs, oyster sauce, taramasalata, roe and, of course, fish themselves, can pop up in the unlikeliest of places!

64

BEERS AND WINES

There are a surprising amount of animal products used in alcohol-making processes. Isinglass is a gelatine-like derivative obtained from fish which is added to real ale to make it more attractive, and used as a "fining" agent in wine, as is gelatine itself. Milk products and even eggs also appear in some production methods. If a beer, wine or cider doesn't state that it is suitable for vegetarians on the packaging then that probably means it has come into contact with animal products along the way. The use of animal products in alcohol production is still not common knowledge, and you may not be able to get an easy answer if you ask bar staff about vegetarian or vegan drinking options. It's best to do your research before you go out and know which beverages you are happy to drink.

65

CHECK FOR CARMINE

The red colouring carmine, also known as cochineal or E120, is derived from crushed cochineal insects. It is used to colour diverse products, ranging from clothes to food to cosmetics. Some pharmaceutical companies even use it to colour medicines such as pills and ointments. When you think of food dye you may think of sweet treats and cakes but carmine is also used in alcoholic drinks, cheeses, juices and sauces – so remember always to check the label for that essential phrase "suitable for vegetarians".

66

WATCH YOUR E NUMBERS

If a product doesn't have the label "suitable for vegetarians" on its packaging but you can't see any obvious animal products on the ingredients list, the most likely culprit is an E number or artificial additive. The E number E904 contains shellac (see p.89) and is used as an ingredient in food and medication. E104 may contain gelatine, E120 uses carmine, E161 and E170 are sometimes derived from the shells of marine animals (as is calcium carbonate) and the artificial sweetener glycerol is made from animal fats unless otherwise stated. These are just a few of the artificial additives that contain animal products – for the full list see online. This can be a lot to remember; you could take a photo of this page or the full list online to refer to in the supermarket while you're still learning.

MEAT-FREE MEALS

Now you've learnt about the meat-free things to try, the things to avoid, new techniques and seasonings to spice up your food and all the hidden things to watch out for, it's time to put your new-found knowledge into action! Here are a few essential recipes to get you started. These recipes are all perfect for meat-free beginners – they are designed to be filling, provide essential nutrients and above all, be easy! Once you've discovered a few favourites you can work up a regular meat-free menu or start experimenting with new cuisines.

67) "MINCE" BOLOGNESE

The purists would say this is more of a ragu. Everyone would say it's delicious!

Serves 6

Ingredients: 2 tbsp olive oil · 2 medium onions, diced · 3 carrots, diced · 2 celery sticks, diced · 4 garlic cloves, crushed · 500 g vegetarian mince · 2 x 400 g tins chopped tomatoes · 2 tbsp tomato puree · 2 tbsp mixed Italian herbs · 1 l vegetable stock

Heat the oil in a deep saucepan on a medium heat.

Add the onions, carrots, celery, mince and garlic. Fry for 15 minutes or until everything is soft.

Stir in the remaining ingredients and bring to a boil.

Reduce to simmer for 50 minutes.

Serve with your choice of pasta.

68) CAULIFLOWER WINGS

Cauli wings are the perfect finger food – hot, spicy, battered and healthy (but not too healthy).

Serves 4

Ingredients: 1 head cauliflower · 100 g plain flour · 200 ml ice cold water · 1 tsp paprika · 1 tsp black pepper · 75 ml sriracha sauce · 2 tsp olive oil · 1 tsp soy sauce

Preheat oven to 180° C.

Chop the cauliflower into small florets.

Whisk together the flour, water, paprika and black pepper until smooth and runny. If the batter is too thick, add more water.

Toss the cauliflower in the batter until well coated.

Line an oven tray with baking paper and add the cauliflower. Cook for 10 minutes on each side, until crispy and golden.

Meanwhile, combine olive oil, sriracha and soy sauces in a small saucepan over a medium heat.

Remove cauliflower from oven, toss in hot sauce mixture and return to oven for 5 minutes to heat through.

69) MEAT-FREE CASSOULET

This is cheap and easy comfort food at its finest.

Serves 2

Ingredients: 1 tbsp vegetable oil · 1 onion, finely chopped · 2 cloves garlic, minced · 400 g tin green lentils, drained · 400 g tin chopped tomatoes · 2 tbsp sundried tomato puree · 2 tbsp Italian mixed herbs (optional) · 4 veggie sausages

Heat the oil in a large, heavy bottomed pan and cook the onion on a low to medium heat until softened and translucent.

Add the garlic and fry for a further 2 minutes, then pour in the lentils and tomatoes.

Stir to combine. Add the tomato puree (and if the puree does not contain additional herbs, add the herbs) and simmer.

Meanwhile fry the veggie sausages for 10 minutes then chop into chunks and add to the lentil stew.

Simmer until sauce is thickened, stirring occasionally.

70) HOT AND SOUR NOODLE SOUP

Butternut squash has a delicious subtle flavour and juicy body that makes it the perfect base for many soups and stews. It's easy to grate but you can halve your cooking time by using a food processor.

Serves 4

Ingredients: 1 vegetable stock cube · 800 ml boiling water · 1 butternut squash, neck section only, peeled and grated · 2 cloves of garlic, peeled and roughly chopped · thumb of ginger, roughly chopped · 3 fresh red chillies, deseeded and sliced · 1 teaspoon ground turmeric · 4 spring onions, sliced · 2 tsp peanut butter · 30 g fresh coriander · 1½ tbsp sesame oil · 1½ tbsp soy sauce · 300 g medium rice noodles · 1 x 400 g tin coconut milk · 300 g green beans, halved widthwise · 2 limes, juiced

In a large saucepan crumble the stock cube into your boiling water and stir until dissolved.

Add the butternut squash and bring your mix back to the boil.

In a food processor blitz together the garlic, ginger, chillies, turmeric, onions, peanut butter, coriander, seasame oil and soy sauce to form a paste.

Add the paste and the noodles to the saucepan. Stir so the paste is dissolved and the noodles separate.

Add coconut milk, beans and lime juice. Bring to the boil, stir through and serve.

71) MUSHROOM WELLINGTON

Wellingtons are the perfect centrepiece to a big cosy meal. Serve with potatoes and veg for the perfect Sunday dinner.

Serves 6

Ingredients: Plain flour, for dusting · 1 tbsp dried thyme · 1 500 g block puff pastry · 2 tbsp olive oil, plus extra if needed · 2 garlic cloves, peeled and finely chopped · 4 portabello mushrooms, stalks removed · 100 g spinach · zest ½ lemon · 200 g vegetarian blue cheese, sliced · 1 egg, beaten

Preheat oven to 220° C.

Dust your work surface with flour and dried thyme and roll the pastry into a rectangle, 5 mm thick.

Heat the olive oil and cook the mushrooms whole with the garlic for 4 minutes on each side. Remove the mushrooms from the pan and pat dry.

If your pan is dry add a splash more oil and, on a medium heat, wilt the spinach. Remove from the heat and drain as much liquid as possible. Add to a bowl and stir in the lemon zest.

Line a baking tray with baking paper and add your pastry. In a vertical line along the middle pile the spinach, then the cheese, then the mushrooms. Pull the pastry around the filling and pinch shut.

Stab several air holes in the top of the wellington, brush with the egg and bake for 40 mins or until golden brown.

72) CHINESE LEMON "CHICKEN"

A taste of takeaway in your own kitchen! Pair with your favourite noodle or rice dish for a cheap, easy and healthy alternative to takeaway.

Serves 4 (with rice or noodles)

Ingredients: 500 g soy meat chunks or 4 fillets, sliced · 1 egg, beaten · 50 g plain flour · 100 g cornflour plus 2 tbsp · 50 g caster sugar · 150 ml lemon juice · 200 ml water

Toss the soy meat chunks or sliced fillets in the egg until coated.

Sift together the plain flour with 100 g cornflour and toss veggie chicken in it until coated.

Fill a heavy-bottomed saucepan with oil 1 cm deep and heat until small bubbles form. Fry batches of veggie chicken for 2 mins on each side or until golden brown all over. Pat dry with kitchen towel when removed from oil.

Meanwhile mix remaining cornflour and water and set aside.

In a small saucepan combine the sugar and lemon juice until simmering and stir in the cornflour mixture to thicken.

Toss the cooked veggie chicken in the sauce and serve.

73) "PULLED PORK" SLIDERS

These light bites are perfect to serve at a summer BBQ.

Serves 4

Ingredients: 2 x 400 g tins jackfruit, drained • 250 ml BBQ sauce • 250 ml water • ½ cucumber • 12 small bread rolls

Place the jackfruit, BBQ sauce and water in a medium sized frying pan and simmer on a medium heat for 20 minutes, stirring occasionally.

Meanwhile use a potato peeler to slice the cucumber into ribbons.

Remove the pan from the heat and use a fork to pull the mixture apart into thin strands until it resembles pulled pork.

Open the rolls and heap with the jackfruit. Top each with ribboned cucumber, close the rolls and serve.

74) PEA AND MINT OMELETTE

This recipe is the perfect mix of fresh flavour and substance, and if you have this for breakfast you'll start the day with a nice hit of protein.

Serves 1

Ingredients: 3 medium eggs · 25 g peas · knob of butter or non-dairy substitute · 3 mint leaves, torn

Whisk the eggs together until the yolk is fully combined with the white.

Boil water and pour over the peas so they partially cook (if the peas are frozen, place in a pan and boil for 2 mins).

Melt a knob of butter in a small frying pan and pour the eggs in.

When the omelette is starting to set, scatter the peas over and cook for a further 3 minutes.

Fold omelette over and scatter mint leaves on top to serve.

75) MIXED BEAN BURRITOS

This burrito recipe is a massive cheat – but it's filling, it's tasty and it's mega quick.

Serves 4

Ingredients: 1 tbsp oil · ½ onion, diced · 2 garlic cloves, minced · 1 pack microwavable Mexican spiced rice · 1 x 400 g tin spiced mixed beans · 4 tortilla wraps · 50 g cheese, grated · 30 g coriander leaves, chopped

Heat the oil and fry the onions and garlic on a medium heat for ten minutes, until softened.

Add the rice and the beans and stir to combine. Cook for 10 minutes, until heated through.

Line your wraps with a sprinkling of cheese and coriander and then portion the rice mix into the middle.

Wrap up and eat!

76) REFRIED BEAN BURGERS

Bean burgers are a staple of meat-free cooking. These spicy, tortilla-chip-encrusted takes on the classic bean burger are about to become a regular weekday treat.

Serves 4

Ingredients: 1 x 400 g tin refried beans · 1 small onion, chopped finely · 2 green chillies, chopped finely · 1 tsp smoked paprika · 100 g tortilla chips, crushed

Roughly mix together the beans, onion, chillies and paprika until beans are broken down and the ingredients are well combined.

Divide the mixture into golf-ball-sized portions. Press gently into patties, around 1 cm thick.

Coat in the crushed tortilla chips.

Meanwhile cover the base of a frying pan in 2 mm oil and heat.

Fry the burgers for a couple of minutes on each side, until cooked through and the tortilla-chip coating is golden brown.

77) TROPICAL KALE SMOOTHIE

This smoothie contains healthy veg and leaves – and enough zesty flavour to pack a punch!

Serves 2

Ingredients: 400 g pineapple chopped, or 1 x 400 g tin pineapple · 100 g apple, unpeeled and chopped · 100 g carrot, chopped · Handful curly kale · 2 tbsp chia seeds · 200 ml mango juice

Add all ingredients to a food processor or smoothie maker and blend until fully combined.

Serve with ice.

78) HOMEMADE TEMPEH

Make your own meat substitute with this tempeh recipe. You can buy the tempeh starter online or in specialist stores. This recipe requires a thermometer.

Makes: 1 loaf

Ingredients: 400 g dried soy beans, soaked overnight and rinsed • 4 tbsp apple cider vinegar • 1 tsp tempeh starter

Cover the beans with an inch of water and cook on a medium heat for 30–45 minutes. Add more water if the pan runs dry.

Add the cider vinegar and cook until beans are soft but not falling apart.

Drain the water and cool the beans to 35° C. Add the tempeh starter and stir until well mixed. Pour the bean mixture into a small glass or ceramic "loaf" dish and gently press until level.

Place inside a plastic box and store in a warm, dry place for 48 hours, checking periodically to ensure the mix hasn't spoiled – if your mix is mushy, slimy or smells bad then you will need to dispose of it.

The tempeh is now ready to be eaten cooked or raw. Store in the fridge in an airtight container for up to a week.

79) HOMEMADE SEITAN

This fake meat can be made using mostly store-cupboard ingredients and a little elbow grease.

Makes: 1 batch

Ingredients: 150 g vital wheat gluten • 75 g chickpea (gram) flour • 1 tsp onion powder • 1 tsp garlic powder • 1.6 l vegetable stock • 2 tbsp dark soy sauce

In a large bowl combine your dry ingredients. In a smaller bowl stir together the soy sauce and 200 ml of the vegetable stock.

A little at a time, add the soy sauce mix to the dry ingredients, stirring gently to combine. The resulting mixture should be well-combined and rubbery.

Knead on a lightly floured surface for 5 minutes. Let sit for 15 minutes and knead for a further 5 minutes.

Divide your mix into smaller balls and gently stretch each piece into "cutlets" or chunks, no thicker than 5 mm.

Add the seitan pieces and the remaining stock to a deep saucepan and bring to a simmer. Cover and cook for an hour. Add more water if it runs dry.

Remove the seitan from the broth and cool. Store in the fridge in a sealed tub for up to a week.

KEEPING IT UP

You've made a positive change in your diet and that's no mean feat! It can be hard to make big changes and say no to foods that you previously loved. If you're feeling your motivation flagging then look to this chapter for ways to stay disciplined and reasons to stick to your mission.

80

READ UP

It may help your motivation to view cutting meat out of your diet as a sort of hobby. Keep abreast of the areas that interest you, whether that be environmental concerns or trying new vegetarian dishes. Eco-warriors can find info in the environmental sections of newspapers and online news sites and by following green campaigners. Those who love the foodie side of it can keep up their foodspiration by following companies and bloggers on social media sites; getting ideas for food to try out, learning about new products and supporting fellow non-meat-eaters. And those who have cut meat for health reasons can track their health journey on a variety of apps.

81

WATCH DOCUMENTARIES

Many a person has cut meat from their diet as a result of watching a particularly chilling documentary – and there are plenty of those out there. If you're feeling that the issues of climate change and animal cruelty are small when compared to the issue of finding something interesting for dinner then try upping your resolve by watching a documentary. It will remind you of all the good reasons that you opted to cut your meat intake in the first place.

82

KEEP THINGS FRESH

When we stick to the same old things we can feel as though we're stuck in a bit of a rut. Sometimes the novelty of something new is energizing – and when the novelty is gone then your motivation to continue also flags. If you're highly motivated by novelty then keep your meat-free diet fresh by challenging yourself to try a new recipe every week. Stay up to date with the latest brands' new meat-free alternatives and make it your mission to try out each new product. You could subscribe to food blogs for new recipes or head out to restaurants and see what new dishes are being created by adventurous and talented chefs for inspiration.

83

FORGIVE YOURSELF

You are just a human trying your best – you will do well most of the time but there will inevitably be the odd slip-up. Perhaps you aim to only eat meat once a week but you bought a ham sandwich for lunch without thinking. Maybe you've cut meat altogether but were tempted to nibble on a nugget at a party. That's OK! If you mentally connect "not eating meat" with "feeling guilty" all the time, you're not going to be very motivated to continue. Instead, forgive yourself and start again the next day.

84

TREAT YOURSELF

It's important to remember to reward your good efforts. Pick out milestones such as "a month without meat" or "six weeks without processed meat" and celebrate them. You can even celebrate simply because you're feeling good about your progress – no milestone needed. Buy yourself a little (meat-free) treat, go out for a drink with a friend or binge watch your favourite show instead of doing your chores – however you choose to celebrate is up to you!

85

BE HONEST WITH PEOPLE WHO DON'T GET IT

How many times can you laugh at a good old vegetarian joke? How often can you commiserate with someone as they explain to you all the reasons why *they* could never cut out meat? Mostly, those jokes and conversations are fine – who doesn't enjoy a laugh or a debate with their friends and loved ones. But sometimes you just want to have dinner together without hearing The One About The Vegetarian for the sixteenth time. Be honest with your loved ones if you are feeling like this and they will likely give you a break (at least for a little while!). And let them have their say – you don't have to agree with it or even take it in if you don't want to, but they won't like being ignored.

86

GIVE POSITIVE FEEDBACK

A lot of large companies are making big positive changes to introduce new meat-free products and offer extensive vegetarian options on their menus. This is because companies respond to consumer demand – most are making these changes because they know they will make money if they do and they may lose customers if they don't – so it's helpful to offer feedback and encouragement to continue this trend. If you've spotted that a restaurant has introduced some new meat-free menu options then write to them on social media or on their website and let them know you're pleased to see it. When a supermarket has a new alternative meat on offer, buy it and see how it is – if it's good, write a review!

87

BE MINDFUL OF THE POSITIVES

Cutting meat not only helps the environment, avoids animal cruelty and improves your health – it feels good too! Be mindful of the positive changes you can feel in your body after cutting meat and you'll feel motivated to keep up the good work. If you find yourself feeling more energetic, reinforce it by consciously thinking "I have more energy today thanks to cutting out meat". If you are pleased by how you've slimmed down, take a moment to appreciate your body and how it keeps you healthy and fit without meat.

TAKING IT FURTHER

So you've chosen to reduce your meat intake, and you've made a positive choice for your health, for the environment and for animal welfare. If this dietary decision is working for you and you are feeling energized and motivated to take further steps to reduce your consumption of animal products, then this chapter can help you keep up the good work.

88

EGGS

The egg industry raises ethical questions when it comes to the treatment of animals. Chicks are "sexed" on hatching and the males are killed, often before they are 2 days old, as they cannot produce eggs and it would lower a company's profit margins to rear them. Battery chickens live short lives trapped in a cage so small they can't move, stacked on top of and underneath other chickens. Even "free range" chickens are kept in cramped conditions and rarely see the light of day. As nutritious as eggs are, they are replaceable. You can get the same nutrients from plant-based foods such as pulses and kale. You can even use fruit, moist veggies and flaxseed to replace them in the baking process.

89

MILK

Dairy farms have as large an impact on the environment as meat farms do. As with all animal agriculture, the dairy farming process is water-hungry and produces high greenhouse emissions. Cows produce enormous amounts of waste which is often not disposed of properly, sometimes polluting local water sources. Dairy farming has its own issues with animal cruelty too: dairy cows are repeatedly artificially inseminated to keep them producing milk. Their calves are taken from them seconds after birth and the male calves are killed to make veal or simply because they're not profitable. Why not try a milk alternative – whether soy, almond, coconut, rice, oat or even cashew milk, most of which contain similar levels of calcium and are fortified with other nutrients.

90

CHEESE

As cheese is made from milk it comes under the same dairy farming umbrella and is responsible for the same ethical and environmental issues. As with meat, humans tend to overeat cheese. It contains necessary nutrients such as calcium but is also calorie dense and high in salt. Ideally humans should eat no more than 28 g of cheese a day as part of a balanced diet. If you think you'd struggle to give up cheese, consider using some of the techniques suggested earlier in this book. You could try cheese-free Tuesdays or only eating cheese for one meal a day. Start slow and see how you go.

91

HONEY

Yes, unfortunately there is even controversy surrounding honey. If you've stopped eating meat because you're against animal cruelty then you may want to look into beekeeping practices. Some beekeepers, even the small-time producers and hobbyists, harm their bees, including cutting the queen bee's wings off so that she can't swarm. However, not everyone believes that honey is a harmful animal by-product. Whether you're comfortable with these practices is a personal decision, but if you decide to cut it out of your diet, alternatives to honey include agave nectar, date syrup or rice syrup.

92

LEATHER

Leather is most commonly made from cow hide. Other common sources include sheep, pigs and goats. The global leather industry kills close to a billion animals every year. Some leather is made from the hides of cows killed for meat, so if you've cut meat for animal cruelty or environmental reasons be aware that you are still supporting the same industry if you wear leather. Lots of products are made from leather, including belts, jackets and wallets. An easy replacement is the very affordable fake leather. For a pricier replacement that matches animal leather in durability look into plant leather or material made from cork.

93

FUR

With the increase of textile technology and the influx of lightweight but insulatory synthetic materials, fur has largely dropped out of practical use. We don't need it to keep warm any more! However, fur is still used in the fashion industry and as a statement of luxury and wealth. Almost all commercially produced fur is farmed and the animals – such as rabbit or mink – are reared with the sole purpose of being killed when fully grown. Although many countries, including the UK, recognize that fur farming is unethical and have enforced a ban, fur products can be imported across the world.

94

CRUELTY-FREE

Items ranging from medicine to cleaning products to make-up are tested on animals. If you've cut meat and other animal products from your diet for ethical reasons then you may also want to be aware of products that harm animals in other ways. The beauty industry in particular has come under fire for their use of animal testing and many companies have responded positively. Many make-up brands no longer test any products on animals, including designer brands, high-street brands and indie brands. Always look for the label "cruelty-free" when shopping.

CONCLUSION

Whatever your reasons for reducing your meat consumption, we hope this book has helped you pick up some tips to invigorate your cooking routine and learn more about the impact meat production has on our world. Don't be too hard on yourself if you slip up, and remember that any adjustments to your diet or lifestyle can be hard to get used to, so go at your own pace. Every little helps! And remember: whether you're cutting down on meat or cutting it out entirely, you've taken a step that will positively impact your health, the environment and the creatures that we share this green earth with. We hope you enjoy the journey!

If you're interested in finding out more about our books, find us on Facebook at **Summersdale Publishers** and follow us on Twitter at **@Summersdale**.

www.summersdale.com